THE FUTURE OF TRANSC.

Stephen J. Blank

August 2002

FOREWORD

The new agreements between NATO and Russia and between the United States and Russia create opportunities for strengthening bilateral and multilateral military activities throughout the former Soviet Union. These could embrace all the militaries of the former Soviet Union and not only enhance military security in the Commonwealth of Independent States (CIS), but also foster military-political integration with the West and possibly defense reform in all or at least some of the CIS regimes. Most importantly, Russia is pledged to cooperate in these activities.

This monograph explores the unprecedented opportunities that are now before the United States and recommends actions that the Government and armed forces, especially, but not only the U.S. Army, should undertake to consolidate and extend the newly emerging military partnership and cooperative security regime that are now developing. Because the opportunities being presented to the United States and NATO were never possible before to this degree, the proper way to exploit them will become a subject of debate.

The analysis and recommendations offered here by Dr. Stephen Blank are intended not only to trigger this policy debate but to contribute to it and provide perspectives for all interested parties. In this regard, this monograph continues SSI's mission of enhancing the formulation and analysis of the U.S. Army and the U.S. Government's national security policies in a new strategic environment.

DOUGLAS C. LOVELACE, JR.
Director
Strategic Studies Institute

BIOGRAPHICAL SKETCH OF THE AUTHOR

STEPHEN J. BLANK has served as the Strategic Studies Institute's expert on the Soviet bloc and the post-Soviet world since 1989. Prior to that he was Associate Professor of Soviet Studies at the Center for Aerospace Doctrine, Research, and Education, Maxwell Air Force Base, and taught at the University of Texas, San Antonio, and at the University of California, Riverside. Dr. Blank is the editor of *Imperial Decline: Russia's Changing Position in Asia*, coeditor of *Soviet Military and the Future,* and author of *The Sorcerer as Apprentice: Stalin's Commissariat of Nationalities, 1917-1924*. He has also written many articles and conference papers on Russian, the Commonwealth of Independent States, and Eastern European security issues. Dr. Blank's current research deals with weapons proliferation and the revolution in military affairs, and energy and security in Eurasia. His most recent SSI publications include "The Foundations of Russian Strategic Power and Capabilities," in *Beyond Nunn-Lugar: Curbing the Next Wave of Weapons Proliferation Threats from Russia*, edited by Henry D. Sokolski and Thomas Riisager, April 2002, and *The Transatlantic Security Agenda: A Conference Report and Analysis*, December 2001. Dr. Blank holds a B.A. in History from the University of Pennsylvania, and a M.A. and Ph.D. in History from the University of Chicago.

SUMMARY

The new U.S. and NATO partnerships with Russia offer an enormous opportunity to shape and transform the security environment throughout the former Soviet Union. The Russian government now supports partnership and integration with NATO and the United States, and Russian military effectiveness is in our vital interest. So the time for an expanded program of engagement with the CIS governments, including Russia, and enhanced shaping of the regional security environment is at hand.

These programs can and should take place under both U.S. and NATO auspices. Their overall objective should be the general enhancement of security and stability in troubled zones like the Transcaucasus and Central Asia. They should contribute to the integration of Russia's armed forces with those of the West, as well as the forces' transformation to a new and reformed model of an army that is more attuned to current strategic realities and more accountable, professional, and subject to democratic control. Similar goals can be postulated for the armies in other CIS countries. Both the United States individually and NATO collectively possess the resources and organizational structures to accomplish this transformation, and many of the governments in the CIS support the overall improvement in military capability and security that such programs would bring about.

Not only would these programs create a lasting basis for strategic engagement with critical states in the war on terrorism, they would also enhance those governments' stability against the threat of insurgency backed by foreign or domestic terrorism, restrain Russia's neo-imperial tendencies, expand democratization of CIS defense establishments, and provide an opportunity for restoring consensus within NATO concerning roles and missions abroad as well as defining NATO's future territorial reach.

To this end, this monograph makes the following recommendations. Based on the existing Russian cell at U.S. Central Command (CENTCOM) headquarters in Tampa, Florida, CENTCOM and the Russian General Staff (GS) should establish a permanent liaison and cell that covers not just Afghanistan, but also Central Asia.

- Once the new Collective Security Treaty Organization (CSTO) for the CIS begins, Russia should invite the Pentagon to send its representatives to be a permanent liaison to the new regional command structure and to the existing antiterrorism center. These links should be integrated within an overall coordination cell.

- U.S. and Russian forces should take advantage of the experience of the CSTO and the Central Asian Battalion (CENTRASBAT) to conduct further combined exercises with Transcaspian militaries. These can and should also be conducted under the auspices of the Partnership for Peace (PfP), and Russia should be encouraged to join and take part as an equal member of PfP. These exercises can and should be supplemented by regular seminars and discussions on threat assessment, doctrine, and coordination.

- A special joint training center could be established at Bishkek or Dushanbe (Kyrgyzstan or Tajikistan). Finally, both Washington and NATO should encourage assigning liaison officers with Russia and Transcaspian militaries at various levels, not just at the CENTCOM/GS level, but down to regional units like Russia's 201st division in Tajikistan and Russian border guards there and Russian liaison units with U.S. forces in Uzbekistan, Kyrgyzstan, and Tajikistan.

- Finally, Washington and Moscow should maintain permanent cells and/or liaison with the new Common European Security and Defense Program (CESDP) organization coming into being in Brussels to ensure tripartite coordination among it, Russia, and the United States (as well as NATO).

These U.S. and allied activities will surely contribute to the long-term stabilization of the region which is in our own and our allies' interests whether or not Russia contributes to those programs. They certainly are also in the interests of local governments. Therefore, the new partnerships we have forged in the CIS, including Russia, offer dramatic opportunities for expanded "defence diplomacy" (to use the British term) and security sector reform that can only have a mutually beneficial impact for all concerned parties.

THE FUTURE OF TRANSCASPIAN SECURITY

Introduction.

Recent American and NATO agreements with Russia and deployments to Central Asia and the Transcaucasus create significant opportunities for building a truly cooperative security regime across the former Soviet Union. The idea that individual governments, NATO, and other Western security organizations effectively could play this role with or without Russia is not new. Four years ago I wrote that the many internal and international challenges to Transcaspian security ultimately pointed to NATO's assumption of a critical regulatory role there. Russian analysts also entertained ideas on new cooperation with the European Union's (EU) emerging defense organs in 2000.[1]

The proposal for NATO's preeminence in the Commonwealth of Independent States (CIS) came under immediate fire from those who felt that Russia should enjoy undisturbed and exclusive hegemony in the CIS and/or from analysts who believed that NATO had either outlived its mission or was, as Russian analysts charged, merely an instrument of U.S. military-political power organized to suppress Russian influence and strength. In other words, they argued that NATO was too effective a check on Russian influence for Moscow to tolerate enlargement. Since then the number of premature mourners at NATO's funeral has also multiplied exponentially. Many of these same observers now argue either that expanding NATO's presence in the Transcaspian might not benefit it because expansion unduly provokes Russia or that NATO after September 11, 2001, is essentially finished as an effective security provider. Still others claim that America cannot foster democracy in the CIS or elsewhere because it has not done so in Egypt or Saudi Arabia. Therefore, these critics argue that Russia should receive a sphere of influence and

1

leadership, if not exclusivity in modernizing those areas and states.[2] Many analysts would also likely have argued that even after September 11, U.S.-Russian or Russo-NATO cooperation in Central Asia was only feasible in the long term. While the joint effort in Afghanistan was a necessary first step, Moscow's fears of the West's presence in the CIS would surely impede genuine cooperation with the West on vital security issues there.

This monograph aims to refute those criticisms. The new East-West partnership offers both the United States and NATO manifold opportunities to exercise a positive influence upon and along with Russia and governments in the CIS to enhance security. This is because the criticisms of NATO as an outdated anti-Russian or suddenly toothless institution wholly overlook or underestimate the positive changes that NATO has undergone since the end of the Cold War, and its great utility for transforming the security situation across Eurasia. Those changes offer the U.S. Government and its armed forces and NATO and its component forces an opportunity to extend the positive transformation they have undergone further afield to reduce the chances of another September 11 or an explosion of insurgency and terrorism in Eurasia or other areas adjacent to or vital to European and American security.

By acting in this fashion, the United States, its armed forces, its allies, and their armed forces can all contribute to the lasting integration of Russia into the West, an outcome that prevents it from trying to upset or revise the status quo in Eurasia and that acts as a moderating and democratizing force for reform within Russia's national defense structure. Additionally, the United States and our allies can foster real progress in deepening the kinds of relationships and engagement with CIS militaries that will make them and their governments reliable partners with the United States and/or NATO in the war on terrorism and in potential future contingencies. Also, these transformative military-political activities and the achievement of the desired outcome of stability and integration of Eurasia with

2

the West reduce the likelihood of future outbreaks of terrorism, insurgency, and violence in an area whose importance to the West as a whole, and not only because of energy, has risen steadily in the recent past. Given the opportunities at hand and the strategic benefits to be gained from exploiting them, it is utterly misguided to assert NATO's uselessness and to refrain from employing available policy instruments to achieve these highly desirable objectives.

Why NATO and Washington Should Act in the CIS and Transcaspian.

The aforementioned criticisms of NATO and of U.S. policy overlook or neglect many facts; first, the fact that there is a real basis for cooperation with Russia that is accepted by Moscow as serving its interests, too. In this respect, they are more mindful of traditional or quasi-imperial Russian interests than is the Russian government. After all, in February 2001 (well before the attacks of September 21) Sergei Ivanov, then Secretary of the Security Council, told Lord George Robertson, Secretary-General of NATO, that joint efforts against terrorism might become the basis of NATO-Russia trust and cooperation.[3] Similarly, some Russian analysts advocate programs similar to those outlined below.[4] Neither do these critics consider the visible disaster of Russian-led modernization in Central Asia and the wider CIS. Nor do they ponder the possibility that partnership with the West can give Russia a more legitimate prominence in the region, albeit one tempered by the demands of partnership. As Richard Haass, Director of the State Department's Policy Planning Staff, recently observed,

> Another area for cooperation is Central Asia, where the United States and Russia have a shared interest in the economic reconstruction in Afghanistan, in halting drug and weapon trafficking, and more broadly in promoting stability, moderation, trade, and development. It seems to me that

3

assuring Russia a prominent role in the economic reconstruction of this region could go a long way towards alleviating Moscow's concerns about the growing U.S. military presence there.[5]

Moreover, these critics also ignore the evolution of NATO and other European security organizations towards cooperative security, the acknowledgment of that evolution by both nonmember states and statesmen, the existence of U.S. programs to engage and transform CIS militaries, and the genuine contribution those programs make to security, stability, and eventual democratization.[6] In fact, these new accords with Moscow permit NATO, the EU, and the Organization for Security Cooperation in Europe (OSCE) to realize the potential inherent in their organizational evolution since 1990 and to do so for and with those endangered states who clearly welcome this enhanced attention to conflict prevention and have repeatedly advocated it.[7] At the same time all these organizations now enjoy Russian cooperation and support.

NATO's critics also overlook that many security challenges and threats to the new Transcaspian states originated in Russian policy; e.g., support for separatists, undeviating support of anti-democratic rulers, and Moscow's incessant search for economic, political, and military hegemony over these states that would radically circumscribe their real independence.[8] Critics of a Western security presence in the Transcaspian also overlook that subordinating CIS governments to Russia's exclusive sphere of influence ensures their endless backwardness, proneness to violent conflicts, and the overall continuation of pathological political-economic-military phenomena afflicting them and their neighbors.[9] Therefore Western security organizations should not use the opportunity arising from these agreements to underwrite failed Russian "peace operations" and neo-imperial policies, but rather help Russia and CIS governments move beyond the failed policies and outlooks of the past and towards stabilizing, developmental, and democratizing policies. Under present

4

circumstances, the way to do this is to change the "correlation of forces" in the area.[10] Instead, as leading Russian defense correspondent Alexander Golts writes,

> The only way a couple of dozen countries can plan and carry out long-term military programs is if they all have transparent and public defense budgets. Only democratic institutions can guarantee that a country's armed forces won't try to influence political decisions and draw the whole alliance into some risky undertaking or other. In other words, the NATO conditions are essential for maintaining confidence among the member states.[11]

In fact, given continuing terrorist threats in and around Afghanistan and to the states of Central Asia, as well as the threats connected with Russian or Chinese interests in hegemony there and proliferation concerns since those states' borders are notoriously porous, a professional military presence there greatly enhances everyone's interests by strengthening the ability of those states to defend themselves against these threats. Given the visible weakness of both the state-building process in the CIS and the real military threats to many of those states, multilateral combined activities would have a strongly positive effect. As observers have noted,

> Shoring up the feeble statehood of several Central Asian states is an important anti-terrorist task for the United States, and nothing we can do to this end is as important as training combat-capable armed forces. We began doing this quietly after the IMU incursion [in 1999-author] but the pace and scope of this aid has greatly increased since September 11. We need to see this process through to success. This will take time and it will be facilitated immensely by a local U.S. military presence.[12]

The New Strategic Environment.

Fortunately, today the pressure of admittedly unforeseeable events has validated the original contention. A radical change in Transcaspian security is taking shape.

At the recent Moscow and Rome summits, the United States, NATO, and Russia formally agreed to work towards a cooperative security regime throughout the CIS. They even agreed to discuss joint peace operations and apparently consider a generic concept for them.[13] This includes joint and cooperative endeavors to bring peace to Chechnya, Moldova, and Nagorno-Karabakh, and wage war on terrorism.[14]

Ukraine also seized this opportunity to announce its intention to apply to NATO, signifying diminished Russian resistance to that membership which otherwise would have profoundly transformed Eurasian security against Russian interests.[15] That decision preserves Ukraine's Western option which is ultimately an option for reforming Ukraine's politics and economic policies and strengthens prospects for authentic geopolitical pluralism in the CIS. Thus Kyiv's decision represents an open defeat for Russian neo-imperial pressure. After all, Russia's Ambassador Viktor Chernomyrdin had publicly criticized Ukraine for its neutrality and not so hidden sympathy for the West, openly stating that Russian preferences should limit Ukraine's sovereignty.[16] Kyiv's decision also strengthens organizations like the Georgia, Ukraine, Uzbekistan, Azerbaijan and Moldova (GUUAM) security organization which has faced constant Russian pressure, and directly rebuffs Moscow's public demand that there be no military or defense organization in the former Soviet Union other than the CIS dominated by Russia.[17]

Although Uzbekistan's sudden decision to leave GUUAM in June 2002 because it feels bilateral relations with Washington and other CIS states offer more guarantees of security weakens GUUAM, that does not reduce the overall significance of Ukraine's decision.[18]

Today U.S. and European forces are stationed throughout Central Asia, Georgia, and Azerbaijan. Other CIS governments seek a comparable presence.[19] Yet all this occurs with Moscow's official blessing as it executes its own

6

rapprochement with NATO, largely on NATO's terms. This does not mean that peace has decisively broken out. There are still very strong forces in Moscow, particularly among the Ministries of Foreign Affairs and Defense, who oppose the Western presence, openly announce their intention to limit that presence, and would subject the CIS to Russian hegemony.[20] Some of their plans undoubtedly enjoy President Vladimir Putin's support since he told the Duma that the CIS is the priority region of Russian foreign policy.[21]

Nevertheless, hitherto inconceivable developments are actually materializing. In May 2002, Putin told the signatories of the Tashkent Collective Security Treaty on the occasion of the conversion of that group into a Collective Security Treaty Organization (CSTO) that he approved of joint collaboration of the CSTO, and explicitly of its members, with NATO. He said that,

> We said at our private discussions today that issues of a military-political nature are also more and more frequently on the agenda within the framework of our organization. The same goes for political issues. It means that we are ready for and open to cooperation with our partners in other similar organizations. It means that the Collective Security Treaty [signed at this meeting-author] could be an element in the forming of new security systems in the world, including in contact with the North Atlantic Treaty Organization.[22]

If this vision can be realized, and there are many obstacles to it, not only in Russia, then it would open the way to a truly multilateral cooperative approach to dealing with the daunting problems of Central Asian and Caucasian security. Genuine multinational cooperation and the advent of a cooperative security regime there also generates new chances for multilateral governance, i.e., multilateral participation in shaping commonly accepted rules of military, political, and economic activity, among governments, nongovernmental organizations (NGOs), and popular grass-roots organizations.[23] Given the enormous and comprehensive range of threats to security in the CIS, especially in and around the Caspian, this would be a major step forward.

The U.S. military presence in Central Asia and the new vision of East-West cooperation have already begun to transform the situation. Pipeline projects that were previously derided as infeasible, unprofitable, and impossible dreams due to Russian opposition, are now viable propositions. Kazakstan shows signs of economic growth, the EU, the European Bank for Reconstruction and Development (EBRD), and the International Monetary Fund (IMF) are all active throughout the former Soviet "space," and numerous plans for reviving trade and constructing major infrastructural projects tying Central Asia and the Transcaucasus and Central Europe to Europe, Iran, and India, as well as East Asia, are all afoot.

These trends are transforming the Transcaspian security situation and carry within them beneficial possibilities for local governments, if only because globalizing their contacts with multiple external governments and institutions reduces chances for any one actor to monopolize or control their policies, thereby ruling out new imperial dreams for the foreseeable future. These trends also reflect the beginning of a hopefully irreversible process of integration and globalization that alone can galvanize their backward economies to progress and escape the well-known and formidable domestic challenges to their security.

Global integration, particularly with real organizations that can provide tangible improvement of local conditions and of physical security, is essential but has eluded the region's grasp since 1991. As Western writers have observed, perhaps the most important factor threatening Transcaspian states' long-term security is the absence of institutionalized mechanisms for resolving inevitable conflicts: ecological, economic, political, ethnic, military, internal, or international.[24] Actions to further integrate these states with Europe are essential for building lasting peace because they strike at that defect, perhaps the most intractable aspect of local states' inability to make a regional peace. Given an external mover and impetus progress might well be forthcoming.[25]

Today's global interaction increasingly connects all the disparate parts of the CIS, former Warsaw Pact members, and the Baltic states with Asia, Europe, and America. It generated their increasing involvement with the European, Asian, and Middle Eastern security agendas even before September 11.[26] Apart from the new opportunity to benefit from the EU's, OSCE's, and NATO'S experience, Transcaspian states can gain in other ways from this globalizing trend. Despite their entanglement in foreign rivalries, these foreign connections and the possibility of their expansion through the construction of new transport, communication, and other infrastructures give hope of overcoming one of the most deep-rooted causes of backwardness, namely being landlocked states located far from major trade routes and the inability to compensate for that factor.[27]

Finally, this new trend towards seeking multilateral and cooperative security solutions in the CIS offers great scope for NATO, the EU, and the OSCE. These organizations, especially NATO, are the most effective recent providers of regional security. NATO, contrary to its critics, is a functioning organization whose integrated military command and political leadership have adapted well to the challenges of the post-Cold War world. Thus it remains the case today that,

> NATO's political procedures and practices were unmatched among security institutions in their design for intensive consultation, commitment to consensus, aversion to the appearance of disarray, and concrete capacity for implementation.[28]

NATO's flexibility and adaptiveness permits it to mobilize, organize, and implement operations to enhance or provide security to threatened states. This capability very much includes the PfP that comprises CIS regimes and which has continued to undergo a comprehensive and rigorous updating to meet contemporary needs and challenges.[29] Thus NATO is an eminently attractive and reliable institution from their standpoint. Unlike the CIS or Shanghai Cooperative Organization (SCO), NATO protects its members from the spillover of military hostilities, prevents other countries from either intervening or being drawn into such conflicts, stabilizes the former Soviet bloc through expansion of membership and organizational tasks like PfP, and reduces the former Soviet bloc states' fear of being left alone face-to-face with Russia.[30]

At the same time, and this is necessary in the CIS though rarely welcome, it provides valuable instruction in tempering the domestic politicization of PfP and members' armies, reduces the nationalization of security policy, and opens a path to the democratization of military forces and overall military reform. In this respect, nothing could be further from the Russian military's relationship to its government, society, and other states. The lack of this control and its dangerous consequences were repeatedly borne out by Russia's involvement in numerous internal wars and coups in the 1990s, its continuing threats against Georgia, and its equivocal record in fighting against Central Asian terrorism.[31]

The Need for Western Intervention.

Precisely because of the dearth of effective mechanisms for resolving and terminating conflicts, organizations like the EU and NATO have a golden opportunity to extend their burgeoning capabilities in conflict prevention and crisis management to new areas and to learn from and overcome previous errors.[32] More importantly, these organizations are eagerly developing those capabilities for conflict prevention, crisis management, conflict resolution, and for peace and stability operations. NATO has been as much about creating, consolidating, and now extending a political or military-political order in Europe, with particular reference to civil-military relations as it has been about a common defense against Soviet or subsequent threats to peace and security.[33] Perhaps more than any other security organization NATO fundamentally has transformed its structure, capabilities, and outlook to assume a much more active, and even potentially proactive role in conflict resolution, crisis management, and peace and stability operations beyond its borders.[34]

The benefits provided by NATO naturally pertain first to defense and security. To the extent that states enter NATO's gravitational orbit, it becomes progressively much more difficult for their armed forces and government to launch or stumble into internal ethnic wars, civil wars, foreign wars, and coups d'etat. The negative examples of the Russian and Yugoslav/Serbian armed forces in the 1990s confirm this. NATO not only helps create a democratic political and military order among its members that precludes war among them, its gravitational pull attracts other states while it restrains their militaries, forcing them to build more democratic, transparent, and accountable military and police structures that are more attuned to international standards and accords concerning the use of force.

Paradoxically, the connection to NATO also makes the new states' armed forces more adaptable to the rigors of modern combat and much more proficient in their task. The almost total inflexibility of their inherited Soviet model cannot be adapted to contemporary warfare and must be overcome before real military progress and integration are achievable.[35] Perhaps this is why Putin asked NATO to help reform Russia's armed forces.[36] Certainly cooperation with NATO and integration with it and the EU might allow those organizations to initiate or assist a transformation among Russia's armed forces in relation to Russian politics and the CIS. Similarly, military reform in Russia and the CIS may fairly be seen as a major move with regard to conventional arms control and security, not to mention

11

Russia's own democratization.[37] Furthermore, it is obvious to all observers that military reform in Russia will not take place from within or of its own accord. As if to confirm this lack of an internal impetus for reform and transparency, Russia's Minister of Defense, Sergei Ivanov, ruled out genuine military collaboration with NATO, a decision that means the Russian military seeks to insulate itself from the "infection" of foreign ideas and standards.[38] Although the idea of creating a dedicated group of Russian specialists for efficient bilateral cooperation with NATO, including military and intelligence personnel, won Ivanov's favorable attention, the decision rests with the Foreign Ministry, another notoriously unreformed institution.[39] And to further limit the scope of Russia's cooperation with NATO, Ivanov also said that not only will there be no operational collaboration, but that Russia, by associating itself with NATO, "effectively proposes a format of security cooperation" that is an alternative to military blocs and alliances, a view seconded by Foreign Minister Igor Ivanov (no relation), who wrote that the new relationship between Russia and NATO "facilitates the transformation of NATO itself in a direction which is in the interests of common European security" and constitutes an alternative to enlargement.[40]

Therefore, if there is to be real and internally generated military reform in Russia—and such reform is absolutely essential for Russia to be secure, prosperous, at peace, democratic, and integrated with the West—external pressure from foreign military-political organizations like (but not only) NATO must constantly be applied. Cooperation cannot become an excuse for Russia to cooperate merely when it chooses to do so or to weaken NATO's capacity for action, while it remains unintegrated with Europe and its military forces unreformed and not subject to international controls and standards to which Russia has agreed.

Examples from the CIS.

Georgia also provides an excellent example of what can and must be done. Georgia's present crises owe much to its failure adequately to control its myriad paramilitary formations, support viable regular military forces, or to deal responsibly with separatist movements, often aided by Russia.[41] Thus there is a constant temptation to exploit the willingness of Chechen forces within Georgia to strike at these forces. This temptation, in turn, gives Moscow an excellent pretext for striking at Georgia by bombing it and sending in uninvited "peacekeepers." The U.S. decision to send forces to train and advise Georgian forces on how to deal with terrorists, not separatists as many in Tbilisi would prefer, both defends Georgia from Russian attacks and reduces its temptation to use irregular forces for ethnic or other internal wars. But it also begins the task of imparting some desperately needed professionalism to Georgia's armed forces and of encouraging Georgia to maintain an army that it can afford.[42] Similarly, the U.S. mission to Azerbaijan helps defend it against Iran's threats against energy exploration in the Caspian and Azerbaijan's coastline, while providing security to energy operations, training, and aid to Baku.[43]

Nor is America the only provider of such assistance. Sweden and Finland have equipped and are continuing to equip almost 19 battalions of Baltic forces, and Finland and Sweden are not only equipping these forces but, along with the U.S. Army, are training them at institutions like the Baltic Defense College in Tartu, Estonia.[44] Turkey and U.S. military officer schools are also providing aid, assistance, and training to CIS officers and armies. Other NATO and American institutions like the Marshall Center and U.S. military colleges are deepening their contacts with CIS militaries. Thus other NATO members and institutions too are taking a leading role in strengthening security and deepening their military engagement throughout the CIS, even to the extent of organizing new security arrangements there. This applies in particular to Turkey and the Transcaucasus.[45] These valuable training and advisory missions not only help inculcate Western notions of military order and professionalism, they provide much needed resources and examples for CIS forces, unlike the Russian forces in Chechnya or Tajikistan, and could provide a basis for helping states like Armenia and Azerbaijan to make peace. More professionalization should also foster a trend towards more rational military expenditures and force structures and help restrain them despite the presence of real threats so that nonmilitary sectors of these states' budgets are not wholly starved of resources. This dimension has been overlooked abroad, but there is no doubt that the terrorist insurgencies in Central Asia since 1999 have stimulated much higher military spending by governments who can ill afford to do so but see no other choice.

Similarly, training and advisory missions or the development of multinational forces like the Central Asian Battalion (CENTRASBAT) teach new skills and missions to forces who need to know how to perform them and create a habit of interaction and mutual trust among them. This relationship then facilitates the projection of power when needed, as in the current crisis. This kind of sustained bilateral or multilateral engagement has become a regular attribute of U.S. military strategy and policy and was institutionalized as a priority program during the 1990s in Central Asia and throughout the U.S. Central Command (USCENTCOM).[46] U.S. military officials attribute Uzbekistan's rapid decision to welcome U.S. and then allied forces into its territory to conduct the war on terrorism to the relationships and learning forged during 3 years of U.S.-Uzbek military interaction.[47]

Nor is this an isolated example. Australian analysts attribute that country's successful operation in East Timor that combined deterrence with cooperation to long-term collaboration with U.S. and Indonesian armed forces. The long-term engagement and joint activities with Indonesia's armed forces gave them the means to assess accurately

Australian capabilities, resolve, operations, and policy.[48] Long-term bilateral and multilateral cooperation among militaries increasingly appears to be a prerequisite for successful prosecution of a wide range of missions in both peace and wartime operations.

Therefore a greater NATO presence in the CIS will offer even more opportunities for support in the war against terrorism and possible future campaigns as well as for the professionalization and democratization of military forces and institutions. This also applies to the EU which is not just building a Common European Security and Defense Program (CESDP) but also creating a multilateral police force that can enter conflict zones and provide a basis for the impartial and disinterested administration of justice and provision of law and order.[49] To the extent that it can provide police power and training along lines comparable to NATO, it too can help restore or strengthen order in endangered societies and help depoliticize and yet professionalize the indigenous police forces. Thus the EU and NATO, working on the military and police dimensions of security in the CIS, can provide actual models for potentially threatened societies.

Thus the kinds of programs suggested here truly accord with vital current needs in international security and not only in the CIS. Dylan Hendrickson and Andrzej Karkoszka of SIPRI observe that,

> The international community is seeking to respond in a more integrated manner to the violent conflicts and security problems facing states. Security sector reform is part of an attempt to develop a more coherent framework for reducing the risk that state weakness or failure will lead to disorder and violence. Where states are unable to manage developments within their borders successfully, the conditions are created for disorder and violence that may spill over onto the territory of other states and perhaps ultimately require an international intervention. Restoration of a viable national capacity in the security domain, based on mechanisms that ensure transparency and accountability, is a vital element of

15

the overall effort to strengthen governance. Security sector reform aims to help states enhance the security of their citizens.[50]

Finally, to the extent that Russia welcomes and cooperates with NATO and EU military-police missions in the CIS, its own forces will hopefully become more professional, sensitized to examples of military-police "best practice" in internal conflicts, and less politicized. At any rate, they will be more constrained politically by the possibility of risking harmony with NATO and the West and thus constitute less of a threat to neighboring states or become less willing to support separatists and rivals to existing governments in the CIS than is now the case. Today the absence of democratic control and professionalism among regional militaries threatens many states in the CIS "shatterbelt," not least Russia itself. In Chechnya, unprofessional, brutal, and corrupt behavior remains the norm. Russian forces there not only resist changing their behavior despite explicit rules to the contrary, but they have also habitually threatened to precipitate a crisis inside Russia if "victory" (which can only mean the destruction of locally organized social life) is denied to them.[51]

Russian generals also assert openly that they feel not enough pressure has been placed upon Georgia.[52] Thus they want to widen the Chechen war, a decision that would have unimaginably bad repercussions across Russia and the Caucasus, if not beyond. Moreover, their operations and threats against Georgia since 1999 are inconceivable without support from Moscow. Thus the strong temptation to strike at Georgia, combined with an inbuilt tendency towards military adventurism in Moscow could easily lead to further provocations and actions along the lines of those in 2001-through April 2002 that could trigger another military conflict involving Georgian, Russian, and now U.S. forces.[53] Russia's armed forces also resist implementing agreements with the OSCE to vacate bases in Georgia and Moldova, participate in the drug trafficking through Tajikistan, and provide much of the flood of stolen or

16

corruptly obtained small arms that are a scourge to the entire CIS. In Central Asia they have played a rather dubious role in the war on terrorism.[54] Moreover, Chief of Staff General Anatoly Kvashnin recently conceded that their condition was "worse than critical," and that they can neither defend Russia nor fight terrorism elsewhere.[55] Subjecting these forces to international norms and standards of military conduct would greatly benefit Russia and its neighbors. Thus the military-political integration of Central Asian and Transcaucasian states into NATO strengthens chances for peace within and among them, without which further progress in any dimension would become exceedingly difficult, if not impossible.

The Need for a Multilateral Approach.

The new post-September 11 reality allows us to envision genuine short- and long-term progress in confronting regional threats to individual and state security. One vision for the future offers hope of concerted international action to pacify the region and then to help it build a better future for all of its citizens. Those concerted actions, it should be noted, build, not just on hope, but also upon existing realities.

Western economic interest in the region beyond energy or energy-related projects has long been established. The rising appreciation of the Transcaspian's relevance for international security predates September 11 and grows out of the preexisting U.S. and NATO military cooperation and engagement with Central Asian and Transcaucasian militaries. But that new strategic importance of Central Asia and the CIS in particular is now a decisive shaping factor of international security.[56] That engagement with NATO and individual Western governments itself reflected the earlier formal entry of Central Asia and the Transcaucasus into the broader European security agenda. As John Roper and Peter Van Ham wrote in 1997, "The main reason why the West cannot remain complacent about

Russia's actions is the fact that Russia's 'near abroad' is, in many cases, also democratic Europe's near abroad."[57]

These processes facilitate the intensifying interaction and mutual engagement with Western governments and Russia. They are and will be an essential instrument of further multilateral progress if the Transcaspian is to move from being a zone of war, as much of it now is, to being a zone of peace. Otherwise, the citizens of all the local governments and, indeed, in more distant lands will suffer the consequences of a breakdown in security and the creation of more black holes in the international order that resist any efforts at reconstruction. Analysts now warn that, if a new center for Islamic terrorism and extremism were to emerge, it would happen in a place resembling Afghanistan where a weak or failed government and an indigenous movement ideologically tied to religious extremism coexisted. In the CIS, Georgia, Kyrgyzstan, and Uzbekistan have recently been singled out as places where this scenario could come to pass.[58] Others see Pakistan's decline into authoritarian rule since 1999 as a harbinger of what could be in the Third World and point directly to the breakdown of controls over police, military, intelligence, and terrorist forces, exactly what the program being suggested here aims to counteract.[59]

In fact, many CIS and other Third World regimes could easily become failing or failed states that materialize a truly Hobbesian nightmare vision where man is a wolf to man and where organized social life has broken down, seemingly with no hope of recovery. And while ideological fanaticism is essential for international terrorism to flourish, it is hardly necessary for any one failing or failed state to become a threat to its neighbors and more distant interlocutors. Nonetheless, the urgency of the situation in the southern CIS is compelling and could easily spread.

Every state in the former Soviet Union is subject, albeit in varying degrees, to the pathologies that make for failing states and then spread abroad. Thus the spiraling

18

criminality of Central Asia that also involves large-scale trafficking in narcotics and conventional weapons (mainly small arms), and several attempted cases of nuclear or other proliferations has now spread to take over crime in Russia's Far East.[60] To the degree that these criminal elements and linked groups can gain control there, they will likely spread further into Russia and East Asia.

Therefore the enormous scale and magnitude of the challenges facing Transcaspian governments and societies precludes any one state or agency from even beginning to contemplate acting exclusively unilaterally to improve conditions there. Multilateral activities are the only way to bring about stability, and an enduring basis for peace, development, prosperity, health, environmental reconstruction, and democracy. Even if we adopt the Clinton administration's view that "job number one" is conflict resolution, our responsibilities in this part of the world only begin but do not end there.[61] Multilateral cooperation on a scale comparable to that occurring in the antiterror campaign is probably the only way in which the international community can begin to act to ensure a better future for this region. And there are vast opportunities for this cooperation.

The present cooperative security regime that has grown around European security, arms control, and the war on terror offers both a model for future collaboration and for extending that regime into the CIS. Extending that regime into the Transcaspian zone also offers Europe, the United States, NATO, the EU, and the OSCE important missions for their future operation and cooperation. Certainly the challenge of establishing new, more relevant, and adaptable military missions and force structures adapted to them provides a way to reinvigorate NATO forces and Trans-atlantic military collaboration in the war on terrorism and to act "out of area" through these renovated force structures.[62] Europe's recent example and evolution is relevant and instructive for the CIS, both because it confronts so many factors of actual and potential conflict and because it is now part of the intricate and often competitive security relationships of Europe, the greater Middle East, and South, if not East Asia.[63]

The numerous efforts to involve local governments in these competitive security relationships since 1991 demonstrate the CIS' participation, albeit sometimes unwilling, in these complex relationships that have taken place in the last decade. Not only are Central Asia and the Transcaucasus integral parts of the global war on terror, they are fast becoming pivotal actors in the global energy economy. In other cases, like the Shanghi Cooperation Organization (SCO), Russia and China sought to sweep them into an organization whose purpose and perspective far transcended regional security issues and reflected their global resistance to American policies.[64]

Central Asia and the Transcaucasus cannot remain aloof from those other regions' challengesas signs multiply of the region's vulnerability to trafficking in narcotics, conventional weapons (either small arms or more lethal platforms), weapons of mass destruction (WMD), and even to involvement in potential nuclear rivalries.[65]Local governments' support for a nuclear weapons free zone and for export control regimes signals their leaders' understanding that a multilateral approach that originates with their own decisive involvement with other states is the only way to go in this regard.[66]

But these challenges to regional security, like the porosity of borders and police corruption, as well as large scale availability of WMD stocks, are also traceable to the Soviet period, the manner of the Soviet state's dissolution, and subsequent trends in politics and economics in all the successor states. No vision of future international cooperation among all the relevant actors here is realizable or sustainable without an honest cataloguing and analysis of the multiple crisis factors that challenge security both now and for the foreseeable future. Regular access to and dialogue with EU, NATO, and OSCE mechanisms by both

Russia and other CIS states would facilitate a more open and candid discussion of these challenges and of ways to respond to them. These security challenges comprise both man-made misguided policies and economic-political-military decisions, and structural elements of the economy like factor endowments or geography. The combined force of their interaction in the region's politics, economics, sociological trends, and ecology threatens to become a negative "force multiplier" for the intensification of simultaneous, multiple, and interactive crises throughout the Transcaspian if we cannot arrest and reverse the negative trends that they represent.

That negative, even Hobbesian, vision of an international or at least regional order composed of several failing or failed states and violent pseudo-states like the Palestinian Authority is the alternative for the Transcaspian region's security future. It is one where cooperation is exceedingly difficult, if at all possible. Here war, poverty, and ecological and social destruction prevail over peace, security, and development, and interstate and internal conflicts are the order of the day.

The Transformation of European Security Organizations and the Transcaspian Mission.

The new entente with Russia and its formal embodiment in the new NATO-Russian Council and Russo-American mechanisms for cooperation against terrorism and for peaceful resolution of conflicts in the CIS open up a vast field of activity for both Russia and all of the West's security organizations. While the CIS may not be at the center of these agencies' work anytime soon, its importance to those organizations will grow steadily. Similarly, the opportunities for developing a cooperative security regime in the region with Russia may also grow commensurately with Russia's integration into Europe. Therefore European security organizations' growing concern with the CIS and with the Transcaspian security agenda will affect these organizations' internal workings, interrelationships, and ties to Russia. Certainly such cooperation could go far in overcoming the present situation where experts like Benjamin Lambeth of the Rand Corporation observe that, given the dysfunctionality of the Russian Air Force and implicitly the entire armed forces, cooperation with NATO would be an "operational nightmare."[67] Indeed, as Dana Allin of the International Institute for Strategic Studies observes, this new partnership once again shows Moscow that NATO is not excluding or severing it from Europe.[68]

Specifically, consolidation and extension of the new working relationship among all these organizations and governments should begin to affect not only the Russian and CIS' governments, but also the internal relationships between EU and NATO. This is especially the case as the former strives to encompass more and more military missions and the latter becomes more and more a primarily political and collective security organization. Therefore, the relations between the EU and NATO, as well as Russia's ties to these organizations, will also change. Indeed, the CIS offers many opportunities for the EU to realize its conflict prevention and peace mission vocation, as well as scope for the exercise of the CESDP's operational mandate over a vast field of activity, and to do so without coming into conflict with America and NATO.

Even more importantly, the opening of this enormous new field for coordinated activity by Transatlantic security organizations can help resolve some of the many tensions now plaguing Transatlantic relationships over international security. These tensions encompass widening disparities in both sides' view of the threats to international and regional security and how to deal with them, the role and missions of armed force, and the most expeditious ways of modernizing and using existing military forces. NATO apparently has concluded that "out of area" missions are now on its agenda, but there is no consensus as to how it should conduct those missions and who should do so.[69] Accordingly, the new opportunities for Transatlantic security organizations and new missions for their agents create an opportunity for bridging the gap or at least for devising an agenda by which that gap might be reduced through multilateral and mutual discussion.

The way forward emerges if we rethink contemporary military operations as constituting a revolution in military affairs (RMA) not only in the technological-operational side, but also relate new operations and weaponry to the concurrent fundamental alterations in the international state system and political order. As Australian analyst Alan Ryan noted, mainly with regard to Canberra's operation in East Timor, the RMA can and perhaps should be reconceptualized in political terms. When we do that, the following points become clear. As Ryan observes, technological transformations in warfare are equaled, if not surpassed, by comparable or larger transformations in attitudes concerning the legitimate use of military power and who may employ military power for what ends.

> Military forces are expected to provide a wider range of capabilities, at less cost, than ever before. For advanced Western countries, this, as well as the expectation that conflict be waged with minimal casualties, has meant that combat capability is no longer perceived as a blunt weapon but rather as a rapier. Accordingly, countries that wish to shape the international security environment need to retain broad-spectrum, leading-edge military forces capable of cooperating with other forces at short notice. The security of the emergent multilateral, international states' system is reliant on general concepts of legitimacy that are increasingly preserved by coalition operations. In the absence of a mandatory international government—a utopian notion at present—the international system is defined by the extent to which states are prepared to cooperate.[70]

Ryan's observations are especially relevant to the possibility for the EU, NATO, and their members to devise systematic bilateral and multilateral plans, including Russia, for preventive deployment, crisis prevention, conflict prevention, and conflict resolution, including actual deployments of forces where necessary. They confirm the necessity for coalitional warfare or stability operations which, in turn, presupposes unity of threat assessments and objectives among the coalition partners. East Timor also confirms the necessity for such capabilities. Australian assessments stress that success was not only due to a long and well-established relationship with Indonesia, but also because of the long and even more established relationship with the United States, and the individual capability of the leading power in the coalition, i.e., Australia, to project power and make independent strategic decisions, capabilities only available to solidly established states.[71]

Similarly, the leading partner must possess robust command, control, computers, communications, intelligence, surveillance, and reconnaissance (C4ISR) capabilities, deployable utility forces, robust logistics and lift capabilities, and thus the ability to protect other states' forces.[72] Kvashnin's admission concedes that only the West can play this role in the CIS, and thus it is essential that legitimacy for future operations be established by the forging of a durable and politically sustainable military relationship among CIS regimes, including Russia, NATO, the EU, and all their members.

The dialogues and activities that would occur within this framework, both strategic and political, also enable NATO to convert many current challenges into opportunities for extending and developing its newly acquired capabilities to meet actual threats and to conduct conflict prevention, crisis management, and both combat and stability operations. Since NATO's militaries must adapt to the new strategic environment defined by Ryan and many other analysts, and also adapt to the requirements of enlargement, a reconceptualization of roles and missions is especially timely.[73]

Military adaptation by the new members and prospective ones is proceeding, but with serious difficulties. It is clear that West European states will not meet the fiscal challenges inherent in developing forces that compare with the U.S. military regarding technological sophistication, or capability for force projection abroad. Indeed, NATO Secretary-General Lord Robertson is already trying to tighten and modify NATO's Defense Capabilities Initiative (DCI) to make it more effective and a more achievable program that can also facilitate NATO's adaptation to the needs of the war on terrorism and enhance cooperation with the CESDP.[74]

Rather than endlessly revisiting the same arguments that now divide NATO, we can exploit the changed conception of the use of force suggested by Ryan and the new opportunities in the CIS to provide a way out of this dead end. It is not only a question of Westernizing new members' militaries and of changing the structures and composition of NATO forces, but also of rethinking mission and procurement strategies among members and/or associated states in the CIS. As Ryan suggests, and he is not alone, the RMA applies universally to political questions more than it does to technological responses to changing strategic circumstances.[75] Envisioning the RMA in this way also allows us to overcome differences between America and Europe regarding the future direction of European defense policies. European governments do not see the need for building extra-European power projection capabilities or forces for foreign combat operations to the extent that Washington does. They also advertise an allegedly strongly differentiated view of the threat from endangered, if not failing states. Therefore, perhaps they should take the lead in designing what U.S. analysts are now calling a Stability Operations RMA that would emphasize the integration of military and nonmilitary activities.[76] Among these activities would be the kind of "defence diplomacy" now conducted by European governments and the United States.[77]

While this RMA would entail investment in new technologies, it would not become the capital intensive process that the U.S. version has become. Certainly, there would be development of new and advanced data bases, artificial intelligence, robotics, nonlethal weapons, and an enhanced military-political analytical capability that embraces both civilian and military institutions. All this would aim to create a culture of creativity regarding preventive, if not post-conflict solutions, missions, and operations in the CIS and elsewhere.[78] And it might actually galvanize mutual activity among all the concerned parties dealing with Transcaspian security involving the roles and missions of democratic or democratizing police and military forces in real or potential combat zones. Admittedly, this might come to resemble NATO's dreaded division of labor where the U.S. forces perform combat operations and Europe gets the unglamorous stability operations. But this proposal (and that is all that it is) could trigger a debate that could lead to mutually acceptable solutions among NATO members and with the EU which has, in any case, announced its readiness to conduct these smaller-scale Petersberg missions. A serious debate, rather than the kind of undisciplined polemics that have now become all too common in the transatlantic debate, would benefit everyone and could stimulate fresh thinking about the strategic integration of Russia and the CIS as part of a reshaped Eurasian strategic system and security agenda.

Conclusions.

Karl Deutsch, one of the pioneers of the theory of regional integration and the originator of the concept of a security community, observed that there are four aspects to regional integration: "maintaining peace, attaining greater multipurpose capabilities, accomplishing some specific task, and gaining a new self-image and role identity."[79] To the degree that NATO and other security organizations effectively systematize and expand their "defence diplomacy" and mutual cooperation with Russia and the CIS, they will certainly facilitate accomplishment of all these goals for Russia and its neighbors and establish new capabilities and a new identity for Europe and its security agencies. These activities could also greatly revitalize Transatlantic cooperation while helping to stabilize and integrate the new states to the West. We should not pretend that this is a short-term process for merely a few years. Russia's integration is already in its second decade and a very troubled affair, not least in its military aspects, and the other CIS regimes are clearly some distance behind Russia.

But if we view today's crisis as both challenge and opportunity, it becomes clear that the war precipitated on September 11 presents vibrant new possibilities for governments and their armed forces to forge new and enduring structures of cooperation. This is only achievable on a multilateral scale, given the size of the challenge in the CIS. But multilateralism, using tested and proven institutions like NATO and the OSCE, as well as the EU's nascent defensive capabilities, provides confidence as well as competence, while not excessively alarming the recipients of this pressure for integration.

Multilateral security engagement on military and other issues not only enhances mutual confidence, but hopefully stabilizes the former Soviet Union and galvanizes the Western security organizations to adopt new missions and forge a new strategic consensus. Few initiatives in world affairs offer so much scope for major positive transformation. Yet not many other situations also hold out the high risk that if we squander the present opportunity the result will be unending conflict across an enormous number of states and territories that may be beyond anybody's ability to extinguish anytime soon. The challenge is therefore great, but so is the opportunity. Moreover, as the means of realizing Eurasia's integration are now at hand, thanks to the new partnerships with Russia and CIS members, the time for action is also now.

Recommendations.

NATO and its leading members' governments should undertake serious organizational moves to consolidate and expand possibilities for effective military integration with the Russian military to give it a stake in integration and in the kinds of reform it must undergo to survive and flourish. It appears that a convergence of approaches is taking place among analysts, if not governments, about the general direction that such integrative activities should take. A Russo-American-European task force composed of the U.S. Atlantic Council, the Centre for European Reform in London, and the Russian Institute for U.S. and Canadian Studies (ISKRAN) reported about the new East-West partnership that,

> In all of this, participation of Russia will be essential. If Russia is not to view out-of-area missions or NATO's new strategic concept with suspicion, it must be involved and informed (although not necessarily with a decision-making role). Russia could be encouraged to take a more active role in the Partnership for Peace, especially as the emphasis on Central Asia increases. Russian officers could also be brought into the more technical process of force reviews, exercises, and even joint force planning. One of the most valuable steps the West could take is to integrate Russia into its own efforts at military reform and perhaps provide at least some of the significant assistance that will be required if the Russian forces are to be transformed into effective coalition partners without placing an unsustainable burden on the Russian society and economy.[80]

Similarly, based on the proposals of the Russian analyst Dmitry Trenin of the Carnegie Endowment in Moscow, NATO's resolutions to its members, and our own thoughts, we can propose the following ideas for consideration by NATO and member governments and militaries. At their first meeting, the Defense Ministers of the NATO-Russian Council in Brussels on June 6, 2002, tasked their ambassadors with several missions, among them:

- To consider the operational implications arising out of the terrorist threat to SFOR and KFOR peacekeepers in Bosnia and Kosovo.

- To determine an appropriate timetable for and proceed with a broader assessment of the terrorist threat to the Euro-Atlantic Area, initially concentrating on specific threats to NATO and Russian forces, civilian aircraft, or by civilian aircraft to civilian infrastructural targets, e.g., nuclear power plants. A working group would subsequently draw the necessary operational implications.

- To consider a follow-on organizing conference in Moscow on the military role in combatting terrorism that should focus on concrete possibilities for enhanced cooperation in this field.

- To consider ways to strengthen cooperation in crisis management.

- To create an ad hoc working group on nonproliferation and develop a joint assessment of trends in the proliferation of nuclear, biological, and chemical agents, leading to a structured exchange of views, and then ongoing missile proliferation discussions, and the exploration of intensified practical cooperation from NBC agents.

- To continue developing the relevant sections of the Permanent Joint Council's earlier work program in arms control and confidence-building measures and develop the NATO-Russia Nuclear Experts Consultation Work Plan.

- To exchange views on defense reform including ongoing NATO-Russian Staff talks on defense reform. These exchanges will consist of dialogue and mutual assistance, including possible initiatives on reform techniques and personnel training and economic issues including conversion.

- To consider creating an ad hoc working group on defense reform and arrange a seminar on defense reform at the NATO Defense College in Rome.

- To cooperate in logistics, air transport, and air-to-air refueling, and "as a first step, to agree [to] specific action plans including possible timetables for taking forward cooperation in these areas"; demonstration tests should be discussed with a view to organizing them as soon as possible.

- To monitor the finalization and implementation for a framework document on search and rescue at sea, including submarine crews and to raise mutual confidence in this area.

- To develop specific plans and agree on timetables for implementing the Council's Work Programme regarding training, exercises, and cooperative air space initiative and to pursue implementation as quickly as possible.[81]

These decisions illustrate how the new Council could serve as a forum for reform of the Russian Army, but if they go beyond these first steps to encompass some of the recommendations made above, they will also internationalize that process and diffuse it to the CIS.

Following Trenin's and our own ideas, the steps below may be recommended to Washington, NATO and perhaps to other leading members of the Alliance and the headquarters of the nascent CESDP in Brussels.

- Based on the existing Russian cell at USCENTCOM headquarters in Tampa, CENTCOM and the Russian General Staff (GS) should establish a permanent liaison and cell that covers not just Afghanistan but also Central Asia.

- Once the new CSTO for the CIS gets started, Russia should invite the Pentagon to send its representatives to be a permanent liaison to the new regional command structure and to the existing antiterrorism center. These links should be integrated within an overall coordination cell.

- U.S. and Russian forces should take advantage of the experience of the CSTO and CENTRASBAT to conduct further combined exercises together and with Transcaspian militaries. These can and should also be conducted under the auspices of the PfP, and Russia should be encouraged to join and take part as an equal member of the PFP. These exercises can and should be supplemented by regular seminars and discussion on threat assessment, doctrine, and coordination.

- A special joint training center could be established at Bishkek or Dushanbe (Kyrgyzstan or Tajikistan). Finally, both Washington and NATO should encourage liaison officers with Russia and Transcaspian militaries at various levels, not just at the CENTCOM/GS level but down to regional units like Russia's 201st division in Tajikistan and Russian border guards there and Russian liaison units with U.S. forces in Uzbekistan, Kyrgyzstan, and Tajikistan.

- Finally, Washington and Moscow should maintain permanent cells and/or liaison with the new CESDP organization coming into being in Brussels to ensure tripartite coordination among it, Russia and the United States (as well as NATO).[82]

These recommendations and decisions illustrate how the new Council could serve as a forum for reform of the Russian Army, but if they go beyond these first steps to encompass some of the recommendations made above, they will also internationalize that process and diffuse it to the CIS. The new partnership with Russia, as revealed in the statement above and in other statements by Putin, Sergei Ivanov, and others, underscores how the questions of Russian defense reform and CIS stability are linked, and that the former is necessary if Russia is to be a lasting and meaningful partner in the war on terrorism and if the Transcaspian is to be securely pacified. In short, the democratization of Russian national security agencies and policies and the stabilization of the CIS are linked policies, and we now have the instrument at hand and the means to use it to bring about genuine progress on those goals.

ENDNOTES

1. Stephen Blank, "Every Shark East of Suez: Great Power Interests, Policies, and Tactics in the Transcaspian Energy Wars," *Central Asian Survey*, Vol. XVIII, No. 2, 1999, pp. 149-184; Moscow, *Nezavisimoye Voyennoye Obozreniye*, in Russian, November 17, 2000, *Foreign Broadcast Information Service, Central Eurasia*, (henceforth *FBIS SOV*), November 22, 2000.

2. Anatol Lieven, "The Not So Great Game," *The National Interest*, No. 49, Winter, 1999-2000, pp. 69-80; Anatol Lieven, "Bobbing for Rotten Apples: Geopolitical Agendas in Ukraine and the Western CIS," paper presented to the Project on Systemic Change and International Security in Russia and the New States of Eurasia, Nitze School of Advanced International Studies of Johns Hopkins University, Washington, DC, 2000; Anatol Lieven, "The End of NATO," *Prospect*, December 2001; Richard Sokolsky and Tanya Charlick-Paley, *NATO and Caspian Security: A Mission Too Far?*, Santa Monica, CA: Rand Corporation, 1999; Eugene Rumer, "Fear and Loathing in the 'Stans'," *Christian Science Monitor*, August 2, 2001; Ira Straus, "Wisdom or Temptation in Central Asia?" *The Russia Journal*, February 22-28, 2002.

3. Moscow, *Interfax*, in English, February 20, 2001, *FBIS SOV*, February 20, 2001.

4. The Atlantic Council of the United States, Centre for European Reform, and Institute for the U.S. and Candian Studies of the Russian Academy of Sciences, *The Twain Shall Meet: The Prospects for Russia-West Relations*, Washington, DC, 2002; Nabi Abdullaev, "Experts Say Moscow To Assume Key Role," *St. Petersburg Times*, June

18, 2002; Dmitri Trenin, "A Farewell to the Great Game?" paper presented to the Conference on Russian National Security Policy and the War on Terrorism, Monterey, CA, June 4-5, 2002, cited with permission of the author (a revised version will be published in a forthcoming issue of *European Security*); for an American view, see Clifford A. Kupchan and Charles A. Kupchan, "Central Asia: A Budding Partnership," *Los Angeles Times*, May 19, 2002.

5. Richard N. Haass, "U.S.-Russian Relations in the Post-Post-Cold War World," remarks to RAND Business Leaders Forum, Tenth Plenary Meeting, New York, June 1, 2002, *www.state.gov*.

6. The best account of NATO's evolution in the 1990s is David Yost, *The Transformation of NATO*, Washington, DC: U.S. Institute of Peace, 1998; for recognition of this by foreign statesmen, see the comments of Finnish Defense Minister Jan-Erik Enestam, "NATO, Europe, and Finland," in Tomas Ries, ed., *NATO Tomorrow*, Helsinki: Department of Strategic and Defence Studies, National Defence College, 2000, from the Seminar on "NATO Tomorrow," arranged by the Atlantic Council of Finland and the Department of Strategic and Defence Studies, National Defense College of Finland, Helsinki, March 30-31, 2000, pp. 1-3; see also Francois Heisbourg and Rob De Wijk, "Debate: Is the Fundamental Nature of the Transatlantic Security Relationship Changing"? *NATO Review*, Spring, 2001, pp. 15-19; Lieutenant Colonel (USA) Timothy C. Shea, "Shaping on NATO's Doorstep—U.S.-Ukraine Relations," *Joint Forces Quarterly*, Autumn 2001-Winter 2002, pp. 58-64; Lieutenant Colonel (USAF) James E. DeTemple, "Military Engagement in the South Caucasus," *Joint Forces Quarterly*, Autumn 2001-Winter 2002, pp. 65-71; and for the underestimation of the value of U.S. and Western military programs, see Andrew Bacevich, "Steppes to Empire," *The National Interest*, No. 68, Summer 2002, pp. 39-54.

7. "Statement by His Excellency Giorgi Burduli, First Deputy Foreign Minister of Georgia," EAPC Foreign Minister Meeting, Florence, May 25, 2000, *www.nato.int / docu. / speech / 2000 / s000525p.htm*; De Temple, pp. 65-71.

8. For examples of the consequences of Russian policy, see Dov Lynch, *Russian Peacekeeping Strategies in the CIS: The Cases of Moldova, Georgia, and Tajikistan,* Royal Institute of International Affairs, Russia and Eurasia Programme, London and New York: Macmillan and St. Martin's Press for the Royal Institute of International Affairs, 2000; Dov Lynch, "Euro-Asian Conflicts and Peacekeeping Dilemmas," Dov Lynch and Yelena Kalyuzhnova, eds., *The Euro-Asian World: A Period of Transition*, New York: St. Martin's Press, 2000, pp. 17-22; Charles King, "The Benefits of Ethnic War:

33

Understanding Eurasia's Unrecognized States," *World Politics*, Vol. LIII, No. 4, July 2001, pp. 538-543.

9. S. Frederick Starr, "Russia and the Neighboring Countries," Presentation to the Kennan Roundtable at The Council on Foreign Relations, Washington, DC, January 24, 2001.

10. *The Monitor*, May 29, 2002; Charles Fairbanks, "Being There," *The National Interest*, No. 68, Summer 2002, pp. 39-55.

11. Alexander Golts, "Tough Challenge Ahead as Putin Looks West," *The Russia Journal*, October 5-11, 2001.

12. Fairbanks, p. 47. It should be noted first that once the mission has been achieved, these forces can then be withdrawn, and second, that they need not be a large number of forces to disrupt local societies, merely enough to fulfill their stabilizing mission, one long-enshrined in U.S. military practice.

13. *Text of Joint Declaration, The United States of America and the Russian Federation*, St. Petersburg, May 24, 2002, *www.whitehouse. gov / news / releases / 2002 / 05 / 20020524-2.html*; "NATO-Russia Relations A New Quality: Declaration by Heads of State and Government NATO Member States and the Russian Federation," Rome, May 28, 2002, *www.nato.int / docu / basictxt / b020528.htm*.

14. *Ibid.*

15. Askold Krushelnycky, "Kyiv, in Policy Shift, Seeks NATO Membership," *Radio Free Europe Radio Liberty Magazine*, May 30, 2002.

16. Charles Clover, "Kiev Warned on Neutral Policy," *Financial Times*, July 12, 2001, p. 2.

17. *The Monitor*, November 30, 2001.

18. Vilor Niyazov, "Uzbekistan Leaves Regional Organization," *TASS*, June 14, 2002.

19. Tom Canahuate, "Uzbekistan, U.S. Plan To Expand Military Ties," *DefenseNews.com*, January 25, 2002; "Armenian Officers Will Get Military Training in US," *Dow Jones International News*, June 20, 2002.

20. See Foreign Minister Igor Ivanov's article in Moscow, *Kommersant-Vlast*, in Russian, June 11, 2002, *FBIS SOV,* June 11, 2002; and *Radio Free Europe Radio Liberty Newsline*, May 21, 2002.

21. This emerges openly in Putin's annual report to the Duma, April, 18, 2002, *www.kremlin.ru.*

22. Moscow, *NTV Mir*, in Russian, May 14, 2002, *FBIS SOV*, May 14, 2002. For an early example of Russian views on how this might be arranged, advanced in order to forestall NATO's first round of enlargement, see Moscow, *Nezavisimoye Voyennoye Obozreniye*, June 21-27, 1997, *FBIS SOV*, June 27, 1997.

23. P. J. Simmons and Chantal de Jonge Oudraat, "Introduction," P. J. Simmons and Chantal de Jonge Oudraat, eds., *Managing Global Issues: Lessons Learned*, Washington, DC: Carnegie Endowment for International Peace, 2001, pp. 8-9.

24. Sokolsky and Chadwick-Paley, p. 10

25. S. Neil MacFarlane, "Arms Control and Peace Settlements: the Caucasus," Working Paper, Geneva Centre for Security Policy (GCSP), 1999-2000.

26. Sokolsky and Chadwick-Paley; David Menashri, ed., *Central Asia Meets the Middle East*, London: Frank Cass Publishers, 1998; Stephen Blank, "A Sacred Place Is Never Empty: The External Geopolitics of the TransCaspian," in Jim Colbert, ed., *Natural Resources and National Security: Sources of Conflict & the U.S. Interest*, Washington, DC: Jewish Institute of National Security Affairs, 2001, pp. 123-142; *FBIS* SOV, November 22, 2000.

27. Ricardo Haussmann, "Prisoners of Geography," *Foreign Policy*, January-February, 2000, pp. 44-53.

28. Celeste A. Wallander, "Institutional Assets and Adaptability: NATO After the Cold War," *International Organization*, Vol. LIV, No. 4, Autumn 2000, p. 724.

29. Pauli Jarvenpaa, "NATO and the Partners: A Dynamic Relationship," in Ries, ed., pp. 48-54.

30. Wallander, pp. 705-735; John S. Duffield, "NATO's Functions After the Cold War," *Political Science Quarterly*, Winter, 1994-1995, pp. 763-787; Rebecca R. Moore, "NATO's Mission for the New Millennium: A Value-Based Approach to Building Security," *Contemporary Security Policy*, Vol. XXIII, No. 1, April 2002, pp. 1-34.

31. On the equivocal position of the Russian forces and government towards terrorism, see Ahmed Rashid, *Jihad*, New Haven, CT: Yale University Press, 2001, pp. 172-174, 189-193; Ahmed Rashid, *Taliban:*

Militant Islam, Oil & Fundamentalism in Central Asia, New Haven, CT: Yale University Press, 2000, p. 179; "Putin's Unscrambled Eggs," *The Economist*, March 9, 2002, p. 53.

32. For an assessment of what European security organizations are doing toward this end, see Paul Eavis and Stuart Kefford, "Conflict Prevention and the European Union: A Potential Yet to Be Fully Realized," in Paul Van Tongeren, Hans van de Veen, and Juliette Verhoeven, eds., *Searching for Peace in Europe and Eurasia: An Overview of Conflict Prevention and Peacebuilding Activities,* Boulder, CO: Lynne Rienner Publishers, 2002, pp. 3-14; Richard Caplan, "A New Trusteeship? The International Administration of War-Torn Territories," *Adelphi Papers* No. 341, 2002; "EU Civilian Crisis Management," in Graeme Herd and Jouko Huru, eds., Conflict Studies Research Centre, Royal Military Academy, Camberley, Surrey, 2001; Wolfgang Zellner, "The OSCE: Uniquely Qualified for a Conflict Prevention Role," Van Tongeren, et al., pp. 15-26; Cees Homan, "Interview with Max Van Der Stoel, Former High Commissioner on National Minorities," *Helsinki Monitor*, Vol. XIII, No. 1, 2002, pp. 3-10; Taras Kuzio, "The OSCE and the CIS: Strange Election Bedfellows?" *Radio Free Europe Radio Liberty Newsline*, April 9, 2002.

33. Moore, pp. 1-34; Wallander, pp. 705-735; Duffield, pp. 763-787; Sergio Balanzino, "NATO Since the Cold War," in Ries, ed., pp. 4-10; Glen Segell, "Civil-Military Relations From Westphalia to the European Union," paper presented to the Annual Convention of the International Studies Association, Minneapolis, MN, March 18-21, 1998; Stephen Blank, "Map Reading: NATO's and Russia's Pathways to European Military Integration," *Occasional Papers of the Woodrow Wilson Center*, No. 61, February 2001, also published in *Review of International Affairs*, I, No. 1, pp. 31-52, 2001.

34. *Ibid.,* Yost.

35. Peter Foot, "European Military Education Today," *Baltic Defense Review*, No. 5, 2001, pp. 23-25; Brigadier General Michael H. Clemmesen (Denmark), "Integration of New Alliance Members: The Intellectual-Cultural Dimension," *Defense Analysis*, Vol. XV, No. 3, 1999, pp. 261-272.

36. Judy Dempsey, "Moscow Asks NATO for Help in Restructuring," *Financial Times*, October 26, 2001, p. 2.

37. Jenonne Walker, *Security and Arms Control in Post-Confrontation Europe,* Oxford: Oxford University Press, 1994, pp.

6-7; Daniel Nelson, *Definition, Diagnosis, Therapy—A Civil/Military Critique,"* unpublished paper, 2001; Golts, "Tough Challenge Ahead."

38. "Moscow Will Not Cooperate with NATO on Military Issues: DEFMIN," *Interfax,* May 3, 2002.

39. Conversations with Russian, European, and Canadian analysts, Wilton Park England, March 11-14, 2002; "Press Opportunity and Joint Press Briefing by Chairman of NATO Military Committee Admiral Guido Venturoni and First Deputy Chief of the Armed Forces General Staff General Yuri Baluyevsky," *Official Kremlin International News Broadcast,* May 28, 2002, retrieved from Lexis-Nexis; Olga Semyonova, "A Group of Experts Will Be Formed for Russia's Cooperation With NATO," *RIA Novosti,* June 6, 2002.

40. *FBIS SOV,* June 11, 2002; "Moscow Will Not Cooperate with NATO."

41. *Radio Free Europe Radio Liberty Newsline,* May 29, 2002; John Diedrich, "U.S. Faces Tough Training Mission in the Caucasus," *Christian Science Monitor,* May 30, 2002; Tbilisi, *Iprinda,* in Georgian, November 18, 2000, *FBIS SOV,* November 18, 2000; Moscow, *Izvestiya,* in Russian, November 23, 2000, *FBIS SOV,* November 23, 2000; "The Ruins of an Empire: Will Georgia Ever Emerge From Russia's Shadow?" *Newsweek International,* June 3, 2002.

42. *Ibid.*; Nikolai Sokov, "Instability in the South Caucasus and the War Against Terorrism," *Program on New Approaches to Russian Security,* Center for Strategic and International Studies, Washington, DC, No. 247, 2002; Thomas De Waal, "Into the Georgian Quagmire," *The Moscow Times,* May 24, 2002, p. 8.

43. For details, see Stephen Blank, "U.S. Military in Azerbaijan To Counter Iran," *Central Asia Caucasus Analyst,* April 10, 2002.

44. Remarks by His Excellency Bjorn Von Sydow, Defense Minister of Sweden, Center for Strategic and International Studies, Washington, DC, May 25, 2002.

45. Canahuate; Igor Torbakov, "A New Security Arrangement Takes Shape in the South Caucasus," *Eurasia Insight,* January 29, 2002, *www.eurasianet.org.*

46. Sami G. Hajjar, *U.S. Military Presence in the Gulf: Challenges and Prospects,* Carlisle Barracks, PA: Strategic Studies Institute, U.S. Army War College, 2002, pp. 19-29.

47. Thomas E. Ricks, "An Unprecedented Coalition," *Washington Post Weekly*, October 1-7, 2001, p. 19; Thomas E. Ricks, "Former Soviet Republics Play a Key Role," *Washington Post*, September 23, 2001, p. A24; Thomas E. Ricks and Susan B. Glasser," U.S. Operated Covert Alliance With Uzbekistan," *Washington Post*, October 24, 2001, pp. A1, 24.

48. David Dickens, "Can East Timor Be a Blueprint for Burden Sharing?" *Washington Quarterly*, Vol. XXV, No. 3, Summer 2002, pp. 29-41; David Dickens, "The United Nations in East Timor: Intervention at the Military-Operational Level," *Contemporary Southeast Asia*, Vol. XXIII, No. 2, August 2001, pp. 228-229.

49. Caplan; Herd and Huru; Eavis and Kefford, pp. 3-14.

50. Dylan Hendrickson and Andrezj Karkoszka, "The Challenges of Security Sector Reform," *Relief Web: Highlights from the SIPRI Yearbook 2002, www.reliefweb.int/w/Rwb.nsf/480fa8736b88bbc3c12564f6004c8ad5/ 3adop25f99be*, June 14, 2002.

51. *Nezavisimaya Gazeta*, November 27, 1999, pp. 1-2.

52. Discussions with Russian generals at the Harvard University Executive Program for U.S. and Russian Generals, Cambridge, MA, February 20, 2002.

53. Sokov; De Waal.

54. Rashid.

55. *Radio Free Europe Radio Liberty Newsline*, May 31, 2002; "Russia Plans Military Overhaul," *BBC News,* May 31, 2002.

56. Adam Daniel Rotfeld, "Introduction: Global Security After 11 September 2001, *Relief Web: Highlights from the SIPRI Yearbook 2002, www.reliefweb.int/w/Rwb.nsf/480fa8736b88bbc3c12564f6004c8ad5/ 3adop25f99be*, June 14, 2002.

57. John Roper and Peter Van Ham, "Redefining Russia's Role in Europe," in Vladimir Baranovsky, ed., *Russia and Europe: The Emerging Security Agenda*, Oxford: Oxford University Press for SIPRI, 1997, p. 517.

58. Douglas Frantz, "Around the World, Hints of Afghanistans to me," *New York Times*, May 26, 2002, Section 4, p. 5.

59. Larry Diamond, "Is Pakistan the (Reverse) Wave of the Future?" *Journal of Democracy*, Vol. XI, No. 3, July 2000, pp. 91-106.

60. "Crime Central," *Far Eastern Economic Review*, May 30, 2002, *www.feer.com.*

61. Deputy Secretary of State Strobe Talbott, "A Farewell to Flashman: American Policy in the Caucasus and Central Asia," address at the Johns Hopkins School of Advanced International Studies, Washington, DC, July 21, 1997.

62. Lieutenant Colonel Raymond Millen, *Tweaking NATO: The Case for Integrated Multinational Divisions*, Carlisle Barracks, PA: Strategic Studies Institute, U.S. Army War College, June 2002.

63. Menashri; *FBIS SOV*, November, 22, 2000; Blank, "A Sacred Place is Never Empty," pp. 23-42.

64. Stephen Blank, "From Shanghai to Brussels: The Future of Transcaspian Security," Forthcoming, *Acque & Terre.*

65. For a general assessment of the relevance of weapons proliferation to the Transcaspian, see Gary K. Bertsch, Cassady Craft, Scott A. Jones and Michael Beck, eds., *Crossroads and Conflict: Security and Foreign Policy in the Caucasus and Central Asia*, New York: Routledge, 2000.

66. Charles Fairbanks, C. Richard Nelson, S. Frederick Starr, Kenneth Weisbrode, *Strategic Assessment of Central Eurasia*, Washington, DC: The Atlantic Council of the United States and Central Asia Caucasus Institute of the Nitze School of Advanced International Studies, Johns Hopkins University, 2001, pp. 78-79.

67. Benjamin S. Lambeth, *The Continuing Crisis of Russian Air Power,"* Santa Monica, CA: Rand Corporation, 2001, pp. 22-26.

68. Jeremy Bransten, "NATO: Russia Signs Accord With the Alliance—Is Membership Next?" *Radio Free Europe Radio Liberty*, June 3, 2002.

69. "Special Report: Europe and America: Old Friends and New," *The Economist*, June 1, 2002, pp. 26-27.

70. Alan Ryan, "From Desert Storm to East Timor: Australia, the Pacific, and the 'New Age' Coalition Operations," *Land Warfare Study Centre*, Study Paper 302, 2000, pp. 1-2.

71. Dickens, "Can East Timor Be a Blueprint," pp. 29-40.

72. *Ibid.*

73. Millen.

74. Judy Dempsey, "NATO Considers Role in Fight Against Terrorism," *Financial Times*, June 6, 2002, p. 2.

75. Ryan, pp. 1-2; Dr. Steven Metz and Lieutenant Colonel Raymond Millen (USA), "NATO and the Revolution in Military Affairs," *Allgemeine Schweizerische Militarische Zeitschrift*, June 2002, pp. 14-15.

76. *Ibid*.

77. "New Challenges to Defence Diplomacy," *International Institute for Strategic Studies, Strategic Survey 1999-2000*, Oxford: Oxford University Press, 2000, pp. 38-53.

78. Metz and Millen.

79. Karl Deutsch, *The Analysis of International Relations,* Englewood Cliffs, NJ: Prentice-Hall, 1978, pp. 239-240; Farkhod Tolipov, "Nationalism as a Geopolitical Phenomenon: the Central Asian Case," *Central Asian Survey*, Vol. XX, No. 2, June 2001, p. 192.

80. *The Twain Shall Meet*, pp. 14-15.

81. "Statement: NATO-Russia Council at the Level of Defence Ministers," Brussels, June 6, 2002, *www.nato.int/docu/pr/ 2002/po20606e.htm*.

82. Trenin.